Juan Carlos Nava

Responsible Use and Management of Agrochemicals

Juan Carlos Nava

Responsible Use and Management of Agrochemicals

Read and understand the label

ScienciaScripts

TABLE OF CONTENTS

TABLE OF CONTENTS 1

DEDICATION 2

THANK YOU 3

Introduction 4

Chapter I Agrochemicals 7

Chapter II Precautions 14

Chapter III Responsible use and handling of agrochemicals 19

Chapter IV Management without agrochemicals 29

Chapter V What not to do 36

Bibliography 41

DEDICATION

To Segunda Chirinos and Angel Noel Herrera, examples of friendship and honesty.

THANK YOU

To the Agronomy students who participated in different research and field activities on some of the aspects reflected.

To all the producers who collaborated with their productive units for the realization of different activities related to the responsible use and management of agrochemicals.

Introduction

Agrochemicals are products that require proper handling and a lot of security, irresponsible behavior cannot be allowed during the handling of the same, to avoid harmful consequences, it is necessary that all persons involved are trained and read and understand the label before obtaining the product to be used, following the manufacturer's recommendations, precautions to take during handling and application, recommended doses, storage, among others.

It is necessary to use alternatives such as biological products to replace the use of agrochemicals, with a contribution to knowledge, where changes are generated that are adapted to each zone, improving their performance levels, conserving their processes and taking care of the environment; with results that can be applied to other zones, for the dissemination of the information generated.

On many occasions, the agrochemical to be used is selected according to a subjective perception of the degree of infestation and climatic conditions, by tradition or because the product is known and trusted in terms of its effectiveness, but unfortunately, without taking precautionary measures for the application of these products, with a high level of risk, since the technical recommendations necessary to avoid the negative impact of these products on human health and the environment are not complied with.

The generation of empty agrochemical containers, as a result of the use of agrochemicals, is a process that grows every year. Therefore, a responsible handling of these containers is required, from the moment the product is obtained until the empty container is discarded. The misuse of these containers can generate sources of contamination, with a risk of toxicity not only for the user, but also for the community and the environment. In many cases, empty agrochemical containers are left in the fields, buried, burned or reused. None of these methods is compatible with the care of the environment.

It is of great importance to work with planning as an indispensable tool in all aspects, where a better quality of social and economic life is provided so that producers, companies, organizations, among others, can be more competitive every day, having the ability to detect errors and correct them, incorporating experiences in their processes, taking care of the environment, establishing strategies that allow them to endure in time successfully.

New knowledge and the new realities of the environment require a change of focus, oriented towards optimization in each area, with the use of internal and external resources and environmental protection; such focus aims to meet the current needs of the population without compromising the environment or with consequences such as imbalances in ecological systems, soil degradation, destruction of watersheds and landscapes, generalized pollution, production processes at accelerated rates, inequity in the use of renewable natural resources, deterioration of the human condition, among others.

Work must be done to develop and implement training and education plans for those involved to develop individual and group capacities, achieving a change of attitude in the use and management of agrochemicals, planning and organizing workshops aimed at reducing the environmental impact and preservation of natural resources, where a culture of quality and social and environmental responsibility is promoted.

Promote training on good practices oriented to improve the quality of the final product, aiming at maximizing productivity in a sustainable manner with diversity and intensity adjusted to each environment.

Therefore, biological alternatives for insect control must be used; integrated management of each crop; continue working to achieve sustainable and sustainable development, with the awareness of people to care for the environment and preserve natural resources by promoting a culture of dialogue among all those involved, including communities, families, schools, colleges, universities, among others.

All activities must be planned as a team, with initiative and knowledge of information and communication technologies, taking into account that the strategies must be applied in specific circumstances, making a responsible use and management of agrochemicals. It is necessary to continue working on knowing and using the different natural alternatives that exist and that can be produced in the companies and productive units, before using a certain type of agrochemical, as well as knowing and using other options, such as biological alternatives, natural alternatives, among others.

Did you know that nature has at least one natural alternative for each of the reasons why a certain type of agrochemical exists?

Chapter I Agrochemicals

Agrochemicals are chemical substances, biological agents or a mixture of both, which are used against agents capable of causing damage to humans, animals, plants, food, among others, as well as to help plants in their growth and development, as in the case of fertilizers. Before using an agrochemical it is necessary to get advice, read and understand the label (Photo 1).

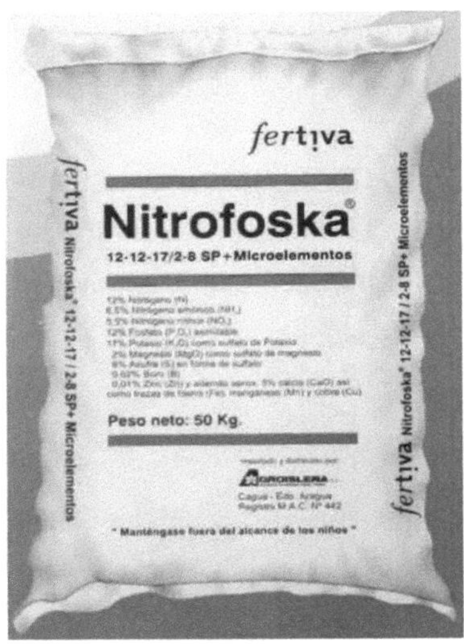

Photo 1. Fertilizer

Formulations

Agrochemicals can be presented in different formulations: solids (powder for dusting, wettable powder, granules, among others), liquids (water-soluble concentrate, emulsifiable concentrate, among others).

Ranking

They can be classified:

- Depending on the type of organism to be controlled: insecticides, acaricides, fungicides, herbicides, nematicides, among others.

- According to the chemical group of the active ingredient: organophosphorus compounds, carbamate compounds, organochlorine compounds, pyrethroids, bipyridyl derivatives, triazines, thiocarbamates, phenoxyacetic acid derivatives, chloronitrophenol derivatives, among others.

- According to their persistence in the environment: Persistent, slightly persistent, non-persistent.

- Acute toxicity (World Health Organization): This is based primarily on oral toxicity in rats and mice. Usually the dose is recorded as the LD50 (Lethal Dose Median) value which is the dose required to kill 50% of the test animal population and is expressed in terms of mg/kg of the animal's body weight.

Insecticide

Chemical, physical or biological agent that eliminates insects or inhibits their growth. They act on insects; there are compounds of vegetable, inorganic, organic and organic-synthetic origin. Example of an insecticide (Photo 2).

Photo 2. Insecticide

Herbicide

Chemical, physical or biological agent used to destroy or inhibit the growth of unwanted plants (weeds). These are products used to control weeds or plants that grow in an unwanted place and avoid competition for water, light, space, nutrients, among others.

They are classified:

• Contact: They are applied to the foliage of weeds, their effect is only to burn the green tissues, they do not move to other parts of the weed.

• Systemic: They are applied to the weeds and if they present mobilization within the weeds, producing a total necrosis of the weed.

- Root-acting: they are applied to the soil so that they are absorbed by the roots and then pass to the upper parts of the plants. Example of a herbicide (Photo 3).

Photo 2. Herbicide

Fungicide

Chemical, physical or biological agent that prevents, inhibits or eliminates fungi. They are products used to prevent or control fungi of inorganic and organic types. They can be contact or systemic with preventive or curative action, applied on foliage and even in the treatment of seeds. Example of a fungicide (Photo 4).

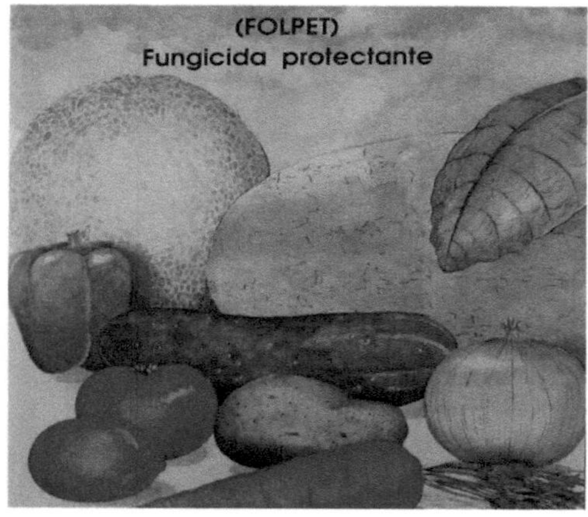

Photo 4. Fungicide

Fertilizers

These are products that help in the growth and development of plants.

They present elements of great importance such as:

Phosphorus

Element little mobile in the soil, its application should be done before planting, at the time of planting or in the first months after planting the plant, it helps in the root development of the plant to achieve the absorption of water, nutrients, organic matter, among others.

Nitrogen

It should be applied fractionally to reduce volatilization losses; it is part of the proteins and the chlorophyll molecule, it is involved in leaf development and plant growth.

Potassium

It is involved in photosynthesis, respiration and fruit development.

Products and Dosage

After taking the soil samples and according to the results of the soil analysis delivered by the respective laboratory, the fertilization program must be prepared by the agronomist who provides technical assistance, placing the product and the dose to be used, after verifying the availability of the fertilizer and its prices in the market. The amount to be applied of a mixture or complete formula is calculated based on its formula in order to maintain the equivalent weights of the individual doses (Photo 5).

Photo 5. Fertilizer

THE AGROCHEMICALS

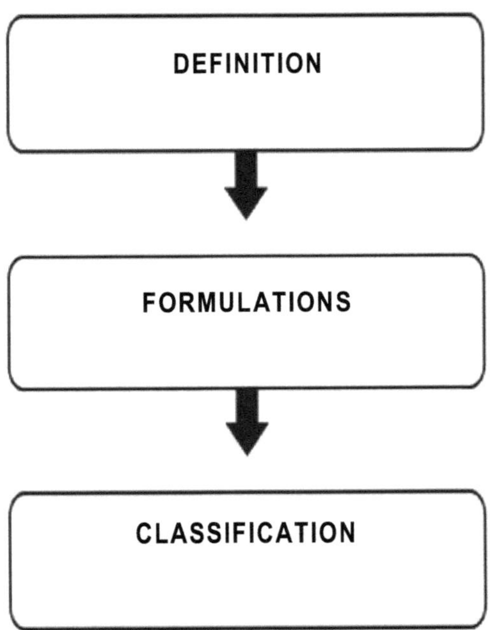

Chapter II Precautions

Most agrochemicals are toxic and can have harmful effects on the body. Agrochemicals can enter the human body by accidental or deliberate ingestion, through the skin, inhalation of small particles or dust during handling or application; exposure through the skin represents the most common risk. There are products that contain the same active ingredient but are distributed under different trade names, therefore, the label must be read and understood before using an agrochemical.

It must be taken into account that it is a function of each State to protect the health of people who carry out all types of activities involving the handling and application of agrochemicals; that a complete medical examination with laboratory analysis prior to exposure to the risk and periodic medical controls are a necessary measure to detect early any alteration in the health of people who carry out these activities.

All persons involved in the use and handling of agrochemicals must undergo an annual medical examination. Persons are considered unfit for the handling and application of agrochemicals:

- Under 18 years of age.

- Pregnant or breastfeeding women.

- People with illnesses.

- Illiterate people.

- People with: Alcoholism, drug dependence, mental disorders, nervous system diseases or diseases that severely affect the respiratory system, skin diseases, cardiovascular system, among others.

No person may enter the activities of handling and application of agrochemicals without a written recommendation from a physician indicating that he/she is in good health.

In this sense, it is necessary that every person who is going to work directly or indirectly with agrochemicals goes to the hospital for a health evaluation, where a medical history will be prepared as a reference document for the following periodic medical examinations during the work activity and will be explained about the risks of handling agrochemicals and about intoxication prevention measures.

The health of all persons involved in the handling and use of agrochemicals should be monitored; the inspection should cover health records and medical examinations, which can alert health authorities to any changes in health that may be related to occupational exposure.

Cholinesterase test

Any person in direct or close contact with agrochemicals should be tested for cholinesterase. Go to the nearest hospital and ask for information to have it done.

It is a blood test that studies the levels of two substances that help the nervous system to function properly. These substances are called acetylcholinesterase and pseudocholinesterase. The person must go to the hospital fasting, a needle is inserted into the vein and the blood is collected; the sample is taken to the laboratory to evaluate these enzymes. These enzymes have the property of breaking down acetylcholine, which is a critical chemical in the transmission of nerve impulses.

Each country may have results with different values and ranges, the important thing is that the day you go to the hospital, the patient should consult with the doctor, the result, if the value is not acceptable, it will indicate a treatment, with the recommendation that should be removed immediately from the work with agrochemicals that performed and a diet where you should not consume alcohol, fats, fried foods, foods such as hamburgers, pizzas, among others.

It is important to emphasize that each organism is different; the Cholinesterase test has been performed on workers in agrochemical distribution companies, and administrative personnel have presented low values and people in charge of product delivery, working directly with agrochemicals have presented acceptable values, so that all people involved in activities with agrochemicals should be tested for Cholinesterase.

If you decide to plan a Monday to go to the hospital to take the Cholinesterase test, the weekend should maintain a healthy diet at home and do not consume alcohol, as the result can be affected, for example, if a person goes to a party on Saturday and consumes fried foods and alcohol and then on Sunday goes out and eats a hamburger, when you go to the hospital on Monday, the result may present unacceptable values, assuming that it is due to agrochemicals.

Environment

The environment must be protected and damage to soil, surface water, groundwater and air must be avoided; the death of organisms that were not intended to be affected must be prevented. Additional problems are arising, such as the accelerated and continuous transfer of residues of these products, chronic exposure of the population to prevent them from consuming contaminated food. Avoid irreversible contamination of soils and well water, avoid causing the death of fish and birds, which alters the ecological balance, and avoid affecting domestic and wild animals.

Crop managers must first evaluate whether the application of the chemical product is warranted; once the decision to apply them is made, a product must be selected that is effective and at the same time consider reducing the hazards for the applicator and the environment. Acquire the agrochemicals in a place intended for this purpose, which should have an agronomist or a technician trained to recommend the best option for the problem to be combated.

The choice and purchase of the equipment to be used for the application of agrochemicals is important. There are several types of spraying equipment on the market; it is necessary to make sure of their condition and operation from a safety point of view in order to take care of the users and the environment. Manufacturers and importers must ensure that the application equipment meets safety and durability standards and comes with its manual and certification.

In this sense, all personnel involved must be trained before handling and applying agrochemicals. The training must be provided by the entity in charge of supervising the handling of the products, with the delivery of a certificate indicating that it covered aspects such as safe handling of products, distribution of the product, use of appropriate application equipment, care of the environment, precautions, among others.

Agrochemicals should only be used if there is an economically important need and strictly according to label recommendations.

PRECAUTIONS

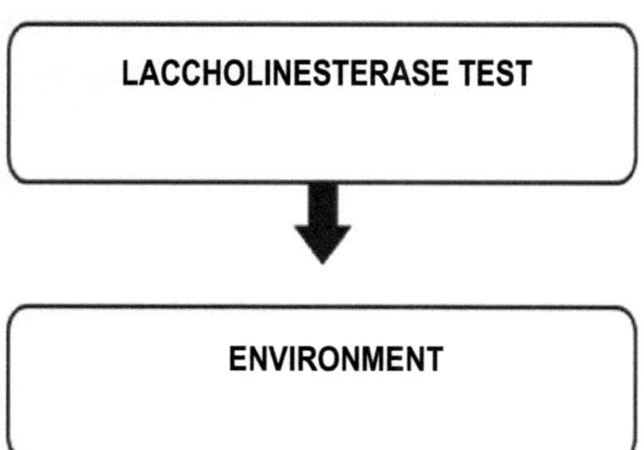

Chapter III Responsible Use and Handling of Agrochemicals

After determining that the application of a particular agrochemical should actually be performed, thanks to technical assistance and having read and understood the label, the following aspects should be covered:

Before the application of an agrochemical must be done:

• Identify the insect, weed, disease, among others.

The tour should be carried out to observe the behavior and plan all the logistics of product application.

• Purchase the product in a safe place, authorized by the competent authority.

Make sure that the distributor is authorized and actually has the product or active ingredient needed.

• Require that the container be original, sealed and with the label in good condition.

The package must not be opened or broken for any reason.

• Verify expiration date.

If the date is out of date, do not buy the product.

• Do not transport the products with people or animals.

Do not bring other people, pets or other animals with the agrochemicals on the day of purchase.

- **Do not drink milk.**

Do not drink milk before applying or handling agrochemicals for any reason.

- Store them away from the home, out of reach of children, UNDER KEY.

Never inside the dwelling, in a room or bathroom; locate a tank and store them in their original container. Under no circumstances should agrochemicals be stored near food.

- Check the application equipment.

That it is in optimal conditions of application.

- Employment of minors is prohibited.

Under no circumstances should minors apply agrochemicals.

During the application of an agrochemical must be:

- Do not eat, drink (no milk) or smoke.

Consult your doctor or the nearest hospital, never drink milk when agrochemicals are applied.

- Apply downwind.

To prevent the product applied by the wind from falling directly on the operator.

- Do not apply if there is rainy weather.

Rainwater washes away the agrochemical and the application is lost.

- Use the triple washing technique; it means to wash the empty container three times, placing clean water up to a quarter of the container, then shake it in different directions for thirty seconds and place the water in the sprayer or tank (Photo 6).

Photo 6. Triple washing

After the application of an agrochemical you must:

- **Do not drink milk.**

Never drink milk after application; consult your doctor or the nearest hospital.

- Perforate the containers.

Make sure that containers or bags cannot be reused; do not burn or bury them; empty and clean containers should be taken to the nearest container collection center.

- Wash hands thoroughly.

With plenty of soap and water.

- Respect the established intervals of re-entry to the treated areas indicated on the label.

The interval period established in the treated area is the time that must elapse between the last application of the product and the entry into the treated field. The pre-harvest interval is the period between the last application of the chemical and the harvest of the crop for consumption. The duration of both periods depends on the agrochemical, the application rate and the crop. The corresponding periods should appear on the label.

- Clean and store the application equipment.

Wash it in a safe place, not in the house, and store it until the next application.

- Separate laundry (Labor).

Do not wash the operator's clothes with those of the rest of the family, always wash them separately.

- Do not use beverage bottles to store agrochemicals.

To prevent intoxication, a person or a child can ingest it thinking it is a drink.

Protective equipment

Nitrile gloves

Indispensable in any treatment and handling of agrochemicals, including the preparation of mixtures, filling, application, washing, among others. They are recommended with Nitrile cover with safety certification for chemical risks. Minimum length of 30 cm and 0.5 mm thick. Once the work is finished, they should be washed. Acids or organic solvents easily damage the gloves. It is therefore necessary to check them thoroughly before putting them on and make sure that they offer the necessary protection. Leather gloves are not suitable for handling chemicals (Photo 7).

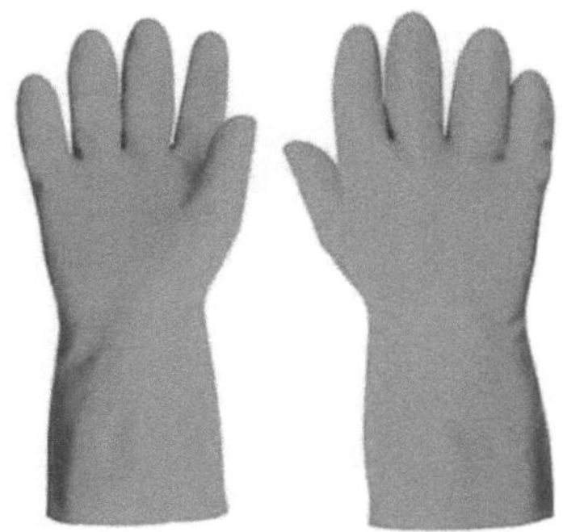

Photo 7. Nitrile gloves

Facial respirators

There are different models, each with different features depending on the product they are working with. The full face respirator offers more protection than the half face respirator; it protects the face and eyes from irritations and contaminants. The important thing is that before buying it, ask if it is for application of agrochemicals, as some are for paint application or for other activities (Photo 8).

Photo 8. Facial respirators

Footwear

Use safety boots or shoes that are in an operative state, with the importance of leaving the pants outside, to avoid that, in case of a leak or spill, the agrochemical can penetrate the socks and then the skin. (Photo 9).

Photo 9. Footwear

Clothing

Wear long pants, long-sleeved shirt and hat, in order to have protection at the time of handling or application (Photo 10).

Photo 10. Clothing

In case of intoxication

- Remove the victim from the accident site.

With great care, to take him to the nearest hospital.

- Take the patient to the nearest hospital with the label or container of the product used.

The hospital will ask what product you were applying.

Initially, in order to purchase the right product, it is necessary to seek technical advice, making sure that the packaging is in perfect condition.

Label

The decision to select a product should be based on an evaluation of the risks and benefits to humans and the environment. In all countries, laws and regulations should be in place to control and regulate the manufacture, import, distribution and sale. Only products that are approved and recommended should be applied. The product label is the first reference for guidance in handling agrochemicals, describing the requirements for use and handling. The label is the main source of information for the user (Photo 11).

Photo 11. Label

The trade name, common name, chemical composition, category, toxicology, storage, formulating company and distributing company appear in the middle of the container label.

On the upper right side are the instructions for use, doses per hectare, which is one of the factors that most influences the success of an application (higher doses may cause phytotoxicity, lower doses may be less effective), crop, control, whether weeds, insects, fungi, among others, preparation of mixtures, compatibility and phytotoxicity. Others such as: re-entry of people and animals to treated areas, last application before harvest, tolerance, precautions, among others, registration, date of preparation and expiration, lot number and net content.

On the left side of the label, there are precautions and warnings for use, symptoms of poisoning, first aid, antidote and medical treatment. Measures for the protection of the environment. At the bottom of the label are pictograms related to storage, handling, recommendations or warnings, being important to observe the illustrations and follow the precautions expressed therein.

In order to avoid fatal effects to the user, precautions should be maintained before, during and after using the products, regardless of whether the product generates a risk of extreme or slight toxicity. Therefore, it is necessary that the user reads and understands the label.

When transporting, place them with the lid upwards to avoid spills; in case of spills, wash contaminated parts and clothing with plenty of soap and water.

Chemical products should be stored in their original containers, well closed, in a special place under lock and key, out of the reach of children, avoiding high temperatures; where they are stored, the use of electrical appliances should be avoided, ventilation should be ensured and a fire extinguisher, sawdust or any other absorbent material should be available (Photo 12).

Photo 12. Fire extinguisher

Labels can change at any time, so it is recommended to read the label every time you buy a product and every time you are going to use it. It is important not to apply in areas not mentioned on the label; there are reasons why some products cannot be applied in different sites, as they may cause damage to certain plant species; present toxicity problems to humans; produce unacceptable residues in crops intended for human consumption, among others.

Do not apply the product inappropriately, read and understand the label; it is not enough to know the type of product; it is also important to know what are the weeds, insects, diseases mentioned on the label and if the product is suitable for the problem you want to treat.

Label instructions indicate which products cannot be mixed and how to test compatibility of products not listed on the label. Do not apply in the wrong location and at the wrong frequency; insects in the soil may require different application sites than those required to control leaf-attacking insects. Respect the maximum number of applications and a minimum interval between applications.

RESPONSIBLE USE AND MANAGEMENT OF AGROCHEMICALS

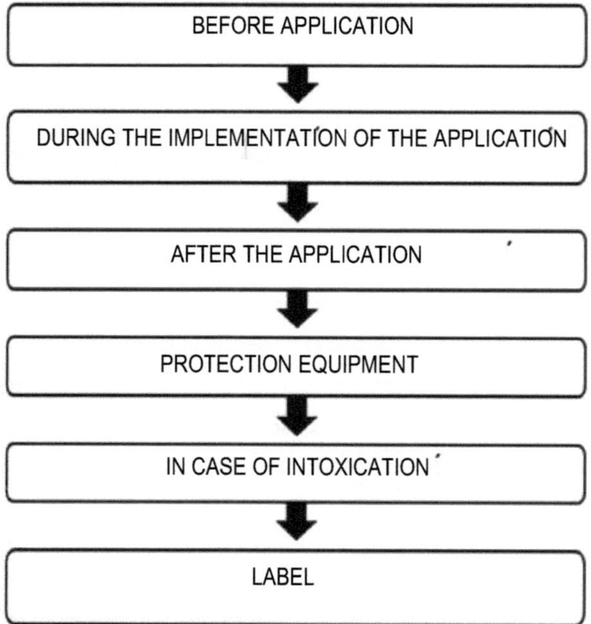

Chapter IV Management without agrochemicals

It is necessary to look for management alternatives that can satisfy the needs presented in the different activities.

Some non-agrochemical management alternatives will be mentioned.

Beauveria bassiana is a deuteromycete fungus that grows naturally in soils all over the world. Its entomopathogenic power makes it capable of parasitizing insects of different species, causing the well-known white muscardine disease. It belongs to the entomopathogenic fungi and is currently used as a biological insecticide or biopesticide controlling a large number of parasites. The mode of action of this entomopathogenic fungus consists of different stages. When the microscopic spores of the fungus come into contact with the cells of the insect epicuticle, they adhere and hydrate. The spores germinate and penetrate the insect cuticle. Once inside, the hyphae grow, destroying the insect's internal structures and causing its death within a few hours. After that, if the environmental conditions are favorable, fungal spores can emerge from the corpse with the ability to be propagated again and re-infect new insects. *Beauveria bassiana* does not contaminate the environment, it is not toxic to humans, plants and animals.

Paecilomyces fumosoroseus is a fungus widely distributed in the soil in most countries of the world that is applied in insect control.

Bacillus thuringiensis is a Gram-positive, soil-dwelling bacterium commonly used as a biological alternative to insecticides. It is a naturally occurring soil bacterium that is fatal to the larvae of a broad spectrum of insects including butterflies, moths, weevils and beetles.

Plant extracts for weed control

Among the disadvantages that arise in the fight against weeds is the irresponsible use of agrochemicals to combat them; these damages affect humans, causing intoxication, because they do not use the necessary protection for their handling.

The inadequate use of agrochemicals also causes problems to the environment, affecting the balance of the soil, generating resistance to the predominant weeds of the place, in addition to the costs that these generate in their acquisition in large quantities, hence the importance of deepening the issue of control of invasive plants through plant extracts, since obtaining the extracts, processing and applying them would be more feasible than buying a product that is an agrochemical and in many cases the responsible use and management of the same is not met.

There are relevant aspects to achieve an integrated weed management, which should be oriented to the integration of different control methods so that they complement each other in terms of ecology and sustainability.

In this sense, there are plants that have an incidence against others, and the weed that predominates in a certain area is able to control the others, so the effect of plant extracts on weed control should be evaluated.

It is recommended to make a tour of the area determined with the agronomist to determine the predominant weeds. The weed that is present in the highest percentage per lot is selected; for example, Cyperus rotundus; it is removed and cut into small pieces, then left to dry at room temperature for 12 hours. For the preparation of the extract, it is cut into pieces of 3 cm and immersed in containers with water for 12 hours, then they are liquefied with the same water that was stored, water is added until a 1:1 ratio is obtained (for each liter of water, 1 kg of leaf); this liquefying is done without spraying (approximately 10 seconds). The preparation obtained is left to stand for 24 hours in covered plastic containers. Subsequently, the liquid is separated from the solid part through a filtering process, placed in the back sprayer and the respective application is carried out; the effect lasts the same as a contact herbicide.

Leachate of banana rachis

Leachate is the liquid obtained from the slow and aerobic decomposition of organic matter. Its importance lies in the fact that it could be a biological controller against *Mycosphaerella fijiensis* and other diseases, acting as a fungistatic agent. The liquid is a "leachate" that is produced economically in the production units and results from the elaboration of compost from banana wastes, specifically from the vertebral axes, called rachis. This is the part of the plant that farmers usually discard after harvesting. To obtain it, in a tank or plastic container, the rachis of the harvested banana bunches, chopped into 10 cm pieces, are placed on a sieve structure that prevents the passage of solid material to the bottom of the tank. After 60 days the final product is obtained. It is placed in the sprayer and the plants are sprayed.

Use of plastic

There is another practice such as the use of plastic (plasticulture), with the purpose of obtaining economic, productive, protection and environmental benefits, improving weed control, reducing water losses due to evaporation, increasing yield and fruit quality. The presence of weeds, especially during the initial growth phase, can cause an appreciable decrease in yields, in addition to the fact that weeds are hosts for insects and diseases.

The use of plastic is an alternative that seeks to obtain higher quality, quantity and mass of fruits, taking care of people and the environment, reducing or avoiding the application of agrochemicals and the use of irrigation water by keeping the soil humidity for a longer time; besides, the land would not be mechanized for several sowings, being convenient for small producers who do not have a tractor and implements and have to rent them (Photo 13).

Photo 13. Use of plastic in the cultivation of melons

Drip irrigation

Use the irrigation system not only to apply water and fertilizers, but also in the application of agrochemicals for controls and thus reduce the use and exposure of people to the products, achieving greater care of the environment, benefiting producers because they would be less exposed to these chemicals.

Fertigation is the simultaneous application of water and nutrients necessary for each crop, a program is made with the daily amount to be supplied. After the drip irrigation system is installed, it is put into operation and the field capacity is evaluated, the amount of water required by the crop, taking into account the type of soil, structure, evaporation, among others. It is measured in time and the decision is made on how many hours or minutes to irrigate per day, with the daily application of the fertilizer or product required by the plants. The fertilization program and other agrochemicals that are proven to be necessary for each stage of the crop are prepared (Photo 14).

Filtering equipment for drip irrigation.

Other alternatives

Use, coffee pulp, crop residues, fertilizers produced by vermiculture, manure, among others, to be included in a fertilization program based on the results of the soil analysis.

Carry out soil conservation practices, such as the application of animal manures, reforestation of forest trees, soil cover, crop residues, organic waste fertilizers, among others; conservation tillage, organic fertilization, recycling of crop residues, cover crops, crop rotation, use of organic products, as an innovative production system that allows obtaining high yields while protecting the environment.

Vermicompost

It is a product of the decomposition of organic matter carried out only by certain species of worms, with a process of decomposition, occurring through their digestive tract and the interaction of certain microorganisms, transforming the organic matter into what is called worm humus or vermicompost, with a better structure and nutrient content than other compost; the liquid obtained during vermicomposting is considered liquid humus (liquid vermicompost). Once obtained, it is placed in the sprayer and the plants are sprayed.

Sulfocalcic Broth

It is a product obtained by boiling a mixture of lime and sulfur; it is a whitish or yellowish orange colored liquid and contains variable concentration of calcium polysulfide. Thirty kilograms of sulfur (ground sulfur flower), ten kilograms of quicklime or slaked lime are added, this dissolved in one hundred liters of water; the water is put to boil and then add the lime and sulfur mixing for an hour, always maintaining the strong intensity in the fire, once the cooking time has elapsed the broth should have a yellowish or orange appearance, it is left to stand until it cools and finally it is filtered and packaged. It is placed in the sprayer and the plants are sprayed.

Hot pepper (*Capsicum sp.* Family: *Solanaceae*).

There are a large number of varieties of the crop that differ in shape, size, color, flavor and spiciness of the fruit, is a valuable food, being an essential ingredient to give flavor and strength to various dishes in the kitchen.

Properties and applications

It has antiviral, insecticidal and repellent action, it is used to control mites, aphids, ants and other organisms that affect the plant; it is considered an option for insect control; the main source of distribution of the insecticide is in the fruit, this being the most commonly used part of the plant.

Add 100 g of ground dried aji to one liter of water and filter with previous agitation. This solution is then diluted in 5 liters of soapy water (20 g of neutral soap) and applied on the crop.

Did you know that nature has at least one natural alternative for each of the reasons why there is a certain type of agrochemical?

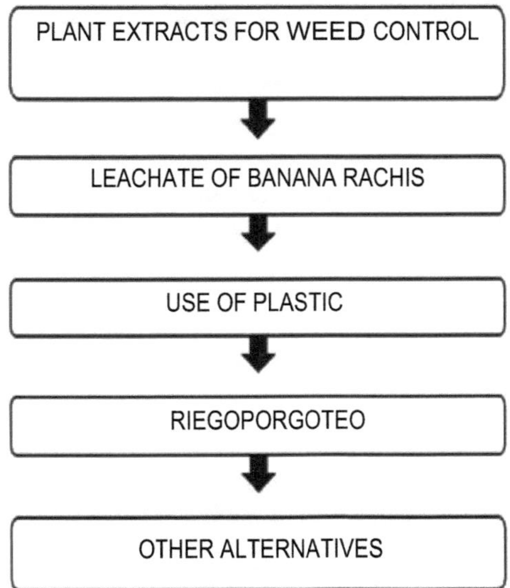

PLANT EXTRACTS FOR WEED CONTROL
LEACHATE OF BANANA RACHIS
USE OF PLASTIC
RIEGOPORGOTEO
OTHER ALTERNATIVES

Chapter V What not to do

If one of the applicators gets dizzy or feels sick, the other can help him/her and remove him/her from the application area.

Photo 16. At the end of the use of the plastic, **do not** leave it in an inappropriate place.

Photo 17. **Do not** leave agrochemical containers anywhere.

Photo 18. Agrochemicals should be properly arranged, they should **NOT be** badly placed and on the ground.

Photo 19. **Do not** employ minors in activities related to the use and handling of agrochemicals.

Photo 20. **Do not** handle agrochemicals without protective equipment, nor near the water to be consumed.

Photo 21. **Do not** apply in areas that have not been previously supervised or in unsuitable conditions.

Photo 22. **Do not** perform unsupervised applications, especially not on another work partner.

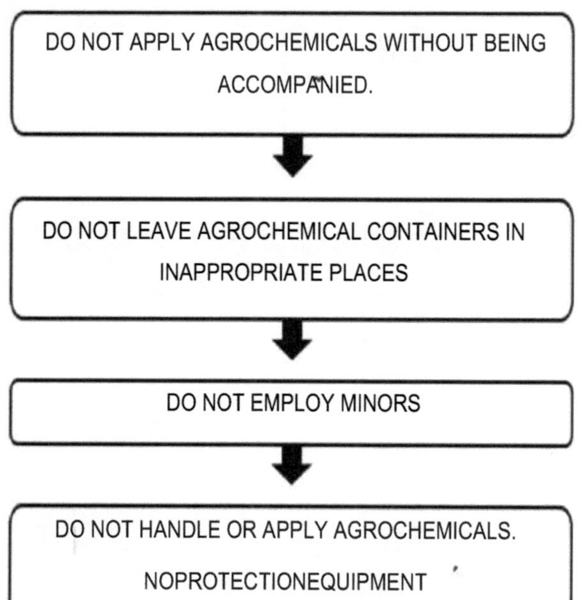

DO NOT APPLY AGROCHEMICALS WITHOUT BEING ACCOMPANIED.
DO NOT LEAVE AGROCHEMICAL CONTAINERS IN INAPPROPRIATE PLACES
DO NOT EMPLOY MINORS
DO NOT HANDLE OR APPLY AGROCHEMICALS.
NOPROTECTIONEQUIPMENT

Bibliography

AFAQUIMA. 2016. Guide for the responsible use and management of pesticides. Association of manufacturers of agricultural chemical products. 20 p.

Alvarado D. 2015 Alternatives for the control of Black Sigatoka (*Mycosphaerella fijiensis*) in banana *(Musa sp.)* cultivation in the area south of Lake Maracaibo. University of Zulia. Faculty of Agronomy. Zulia State. Maracaibo - Venezuela. 43 p.

Barboza, A.; Hernàndez, C.: Nùnez, j.; Nava, J. and Gómez, A. 2014. Plant extracts for the protection of the bunch of Cambur Manzano (*Musa AAB*) in the Baralt municipality of Zulia state. University of Zulia. Faculty of Agronomy, Department of Statistics. Chair: Agricultural Research 44 p.

Medrano, C.; W. Gutiérrez and Ortega, S. 2012. Responsible use and management of agricultural pesticides. University of Zulia. Academic Vice Rectorate. Producciones Editoriales C.A. Mérida, Venezuela. 44 p.

Nava, J. 2004. Precautions in the use of agrochemicals. AGROTECNICO. Extension magazine of the Faculty of Agronomy, University of Zulia. No 17. p 34.

Nava, J. 2009. In agricultural extension, planning is a fundamental tool. AGROTECNICO, extension magazine of the Faculty of Agronomy, University of Zulia. No 35. P 17.

Nava, J.; Lois, G.; Molina, Y.; Faria, E. and Gómez, A. 2013. Plant extracts for weed control in banana crop, Zulia state. University of Zulia. Faculty of Agronomy. 20 P.

PACAZ. 2008. Manual of good agricultural practices. Program of technical assistance and agricultural training of Zulia state. Convenio Gobernación del estado Zulia- Universidad del Zulia-SDA Zulia. 109 p.

Rincon, M.; Morales, H. Nava, Juan and Gil, Marcelo. 2014. Sustainable development of the community of Cherepta de La Sierra de Perija. Zulia State. Rev. Fac. Agron (LUZ). 32:381-406.

Printed by Books on Demand GmbH, Norderstedt / Germany